FRONT AND
REAR WING

It is the first book in a large and special series of books, dedicated to motorsport in general; it will cover aerodynamics, suspension, engines, dynamics, etc. ..... Everything you need to learn how to design a full car.

The aim of this series is also to say that I would like to teach again in a university.

I hope that this series will be a success and that I will be able to transmit all my knowledge and all my experience.

@TimoteoBriet

WINGS

First, we must consider the issues of designing a functional wing.
There is a saying in engineering, which says:

*"If a system serves several
things the better."*

This means that even though a certain system serves mainly for a goal and is the reason for its existence, we must try and use it for other purposes that are aside but that could improve the car's aerodynamics.

The goal of placing a wing in a racecar on the first place is to generate downforce; in fact, at the beginning of the racing history, wings were not used for anything else (Wikipedia images):

Wings may be of two types:
   1) Formula: type wing

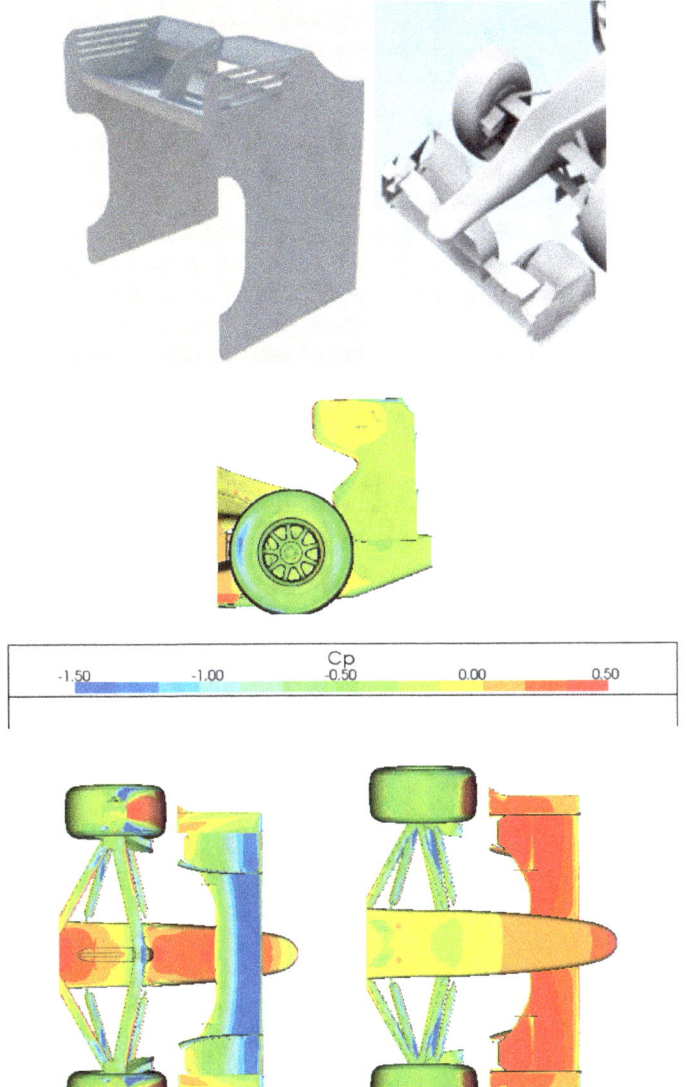

2) Dive planes or flicks:

In both cases, their function is to generate downforce.

The simplest wing is called dive planes. They are normally installed in closed wheel competition cars where the engine is located at the rear so a bigger percentage of the total weight will be placed in the rear axle. This causes understeer making it necessary to increase the front loading using these devices.

The rules where these devices are allowed are clear and they generally say:

It is possible to place any aerodynamic element in that area, provided that it does not protrude into the contour of the car's floor.

For this reason, there are cars where dive planes are work great and perhaps very efficiently, and others, where the efficiency due to their small size and critical position is not so great.

However, the main objective is to generate downforce at the front; we know that this will entail a change in the rear load; later we will see that they can have an important role in increasing the ground efficiency.

FRONT WING

FRONT WING- 1

A F1 type front wing has many functions:

a) To generate downforce.
b) To redirect the flow for cooling.
c) To adjust the flow passing underneath the car.
d) To adjust the flow with the rest of the car.
e) To redirect the flow going to the wheels.
f) To support other aerodynamic elements such as spoilers or vortex generators to reduce vibrations.

Each season rules change to improve safety: the size and position of the front wing changes; It is perhaps the most important aerodynamic device as it is the first device that gets in contact with air. Any small change completely alters the behavior of the rest of the car.

Ground clearance is of great importance to the variation of load generated because we know that viscosity is very important in the boundary layer:

We can quantify the influence of downforce and drag of a profile with chord "c" and ground clearance "h":

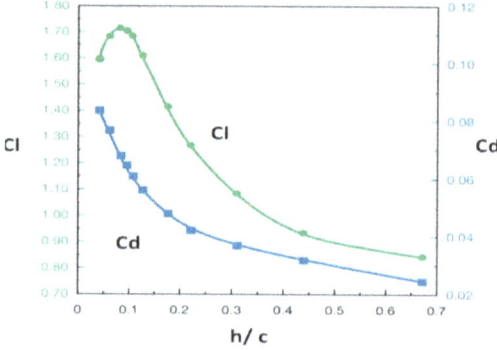

The bunny ears and fox ears, have the capability of increasing the attack angle of front wing, and improve the floor effect underneath car. Also are very important for stability and cross wind (vortex creation):

The smaller the ground clearance, the higher the load generated up to a limit; this limit is imposed by the viscosity of air. There is a moment where there is a layer of high density due to viscosity that prevents the passage of air and therefore stops generating downforce. It is important to know this limit; In any case this limit is located at a very low distance from the ground. This limit is not calculated using only the Bernoulli energy equation. Let's take a look at a CFD simulation (blue color → less pressure):

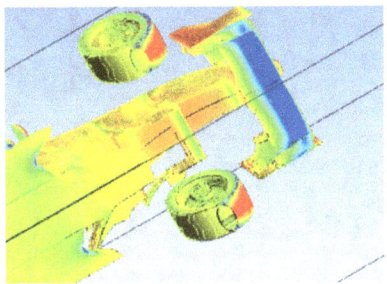

In relation to the unusual "C" shape that F1 front wings have, we must say rules say nothing about it. This shape is due to the existence of sidepods; without this shape the front wing's high angle of attack would prevent airflow reaching the sidepods. This air is very necessary for cooling, so it could mean a complete failure of the powertrain system.

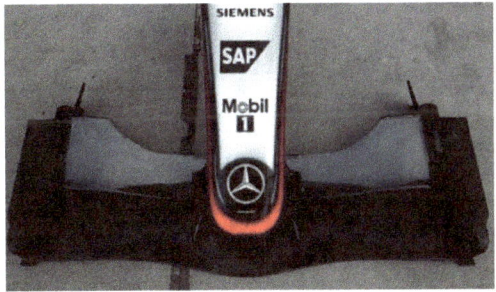

The central part (neutral part), is the same for all teams (placed by FIA):

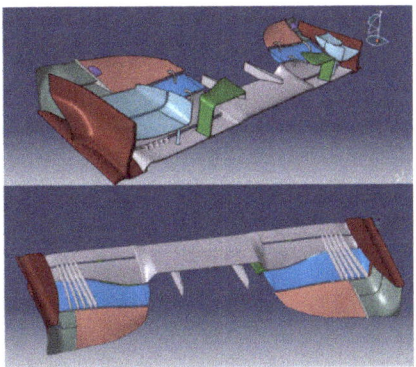

Also, produce vortex (we will see that in this chapter and the next chapter):

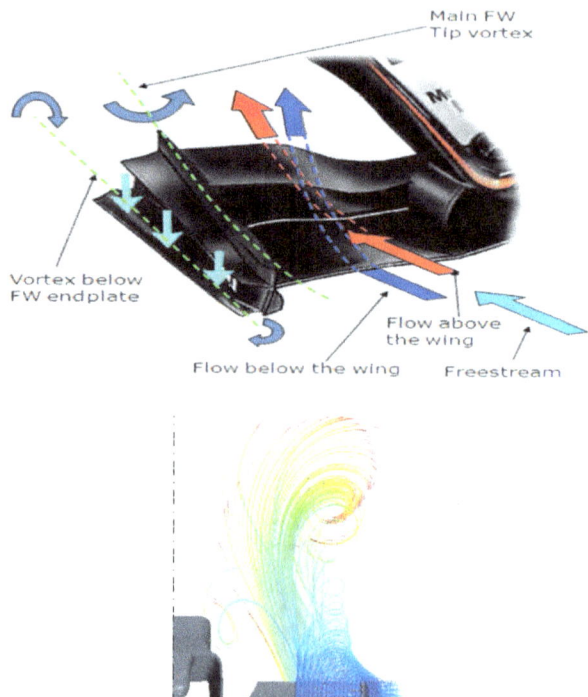

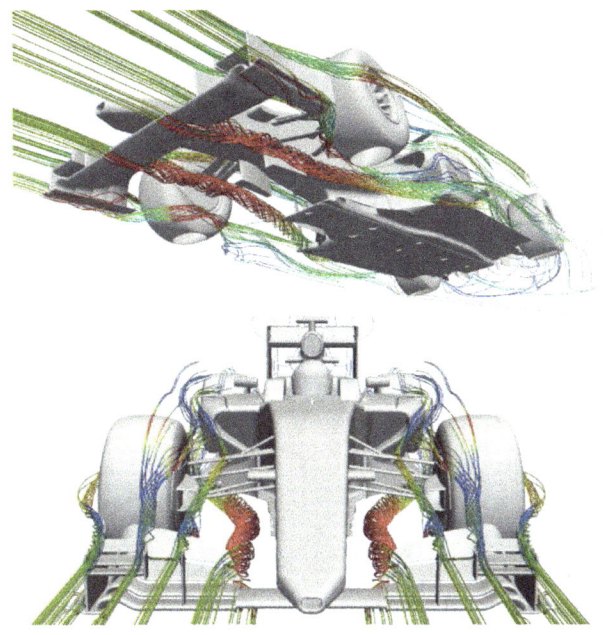

Y250 Vortex:

The Y250 vortex, similarly to wing tip vortices is generated because of a pressure difference, in this case between the neutral middle section (250mm from centerline) and the rest of the wing. It is very important for controlling flow approaching leading edge of the floor. Here in analogy with the delta wing vortex lift, the Y250 vortex can potentially extract air at the edge of the floor, therefore producing a suction effect that improves aerodynamics efficiency in this area. On the other hand, the Y250 vortex may also have a benefit on managing front wheel wake by pushing it away from the car.

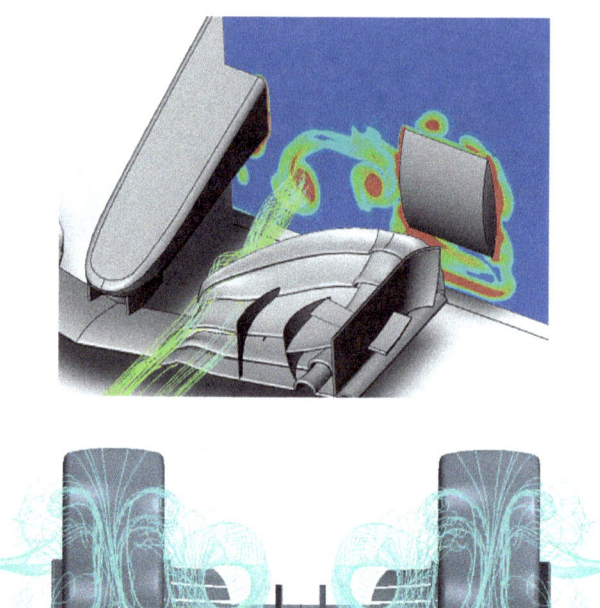

This vortex is controlled and amplified by the down part of sidepod, in order to direct it to diffuser or rear part in general; more, this Y250 vortex, is an artificial skirt along side car:

Also controlled and amplified by the "bat wing":

This bat-wing, produce:

- Low pressure up wing.
- High pressure down wing.

These both pressure can adapt the air flow and/or vortex Y250, for example.

Skysports has edited a good video comparing Y250 coming of RBR and Ferrari and talking about how well-controlled the RBR vortices is. However 'well-controlled' is never a good word for engineers to use. To estimate vortices, we need to use some parameters like vortex core position, vortex strength and cleanness.

These parameters are what the design of complex flaps/cascades is based on. I can't agree with the Skysports' explanation of vortices travelling down over sidepods as the vortex strength cannot be strong enough to affect area so far downstream.

Nevertheless the Y250 vortex do interact with different parts of the car, which requires careful consideration in designing of front wing, floor and turning vanes, etc:

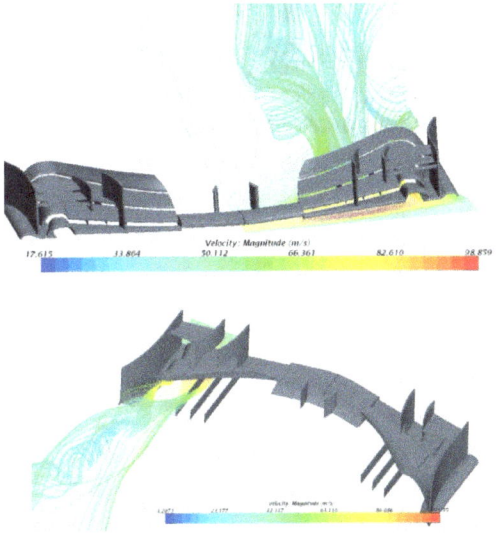

This vortex Y-250, is necessary to control from low and high pressure zones, in order to allow the rear zone, also as a skirt:

Here we can take a look at some important considerations about the front wing:

- Generating a lot of downforce is relatively easy; what happens is that you are altering the flow at the back.
- If you are required to maintain the position of the car's pressure center increasing the downforce, it is necessary to adjust both the front and rear spoiler.
- If the front wing angle is adjusted exclusively, the global drag will remain constant. But variating the pressure center....
- If it is required to increase the angle of attack, it is best if we adjust the side flaps; incidentally, we are reducing the total resistance as more flow travels over the wheels.
- Keep in mind that it is the first aerodynamic features of the car. Hence we must take into consideration its importance and especially its impact on the rest of the car.
- If we only vary the rear wing angle, we see that the front wing downforce also varies.

Returning to the points related to the functions of the front wing. In order to design a good front wing we must avoid flow detachment from the surface of the wing, In a CFD simulation when we display the pressure on the surfaces we should be able to see a uniform color; if there is "discontinuity" in color, "something is happening".

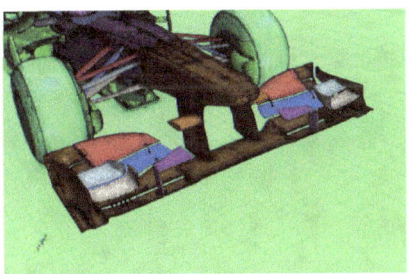

The dark color means more downforce is generated or it is an area of depression: color discontinuities may be due to the transition between two surfaces or due to "abrupt" changes in load or pressure.

The front wing consists of two parts: a fixed part corresponding to the part of the wing that is fixed to the nose and two flaps one on each side, which we can adjust to generate more or less downforce:

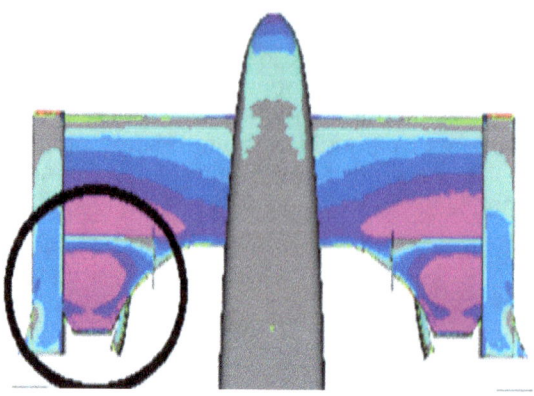

In the image below we can see how the different positions of the screws can vary the angle of attack of the wing (example: F3):

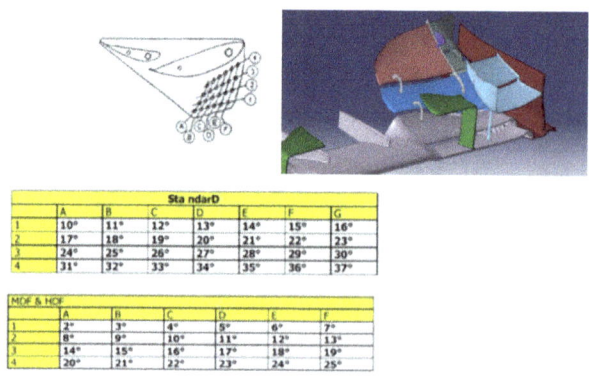

Standard						
	A	B	C	D	E	F	G
1	10°	11°	12°	13°	14°	15°	16°
2	17°	18°	19°	20°	21°	22°	23°
3	24°	25°	26°	27°	28°	29°	30°
4	31°	32°	33°	34°	35°	36°	37°

MDF & HDF					
	A	B	C	D	E	F
1	2°	3°	4°	5°	6°	7°
2	8°	9°	10°	11°	12°	13°
3	14°	15°	16°	17°	18°	19°
4	20°	21°	22°	23°	24°	25°

The front wing can also be used for other things that may look distant or totally unrelated to the main goal, which is to generate downforce: McLaren wings produces an increase in engine power of about 7 hp, due to suitable channeling of air flow to the engine inlet:

FRONT WING – 2

There are other front wings, called dive planes, which are responsible for generating front downforce; obviously they also generate drag and vary the center of pressure; besides generating downforce they are also responsible for the generation of longitudinal vortices whose function is to increase the efficiency of the ground effect (skirt) (is very complicate to generate that):

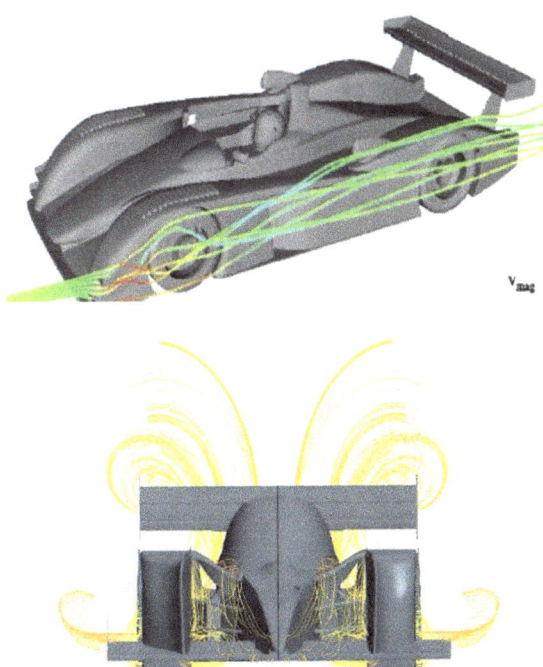

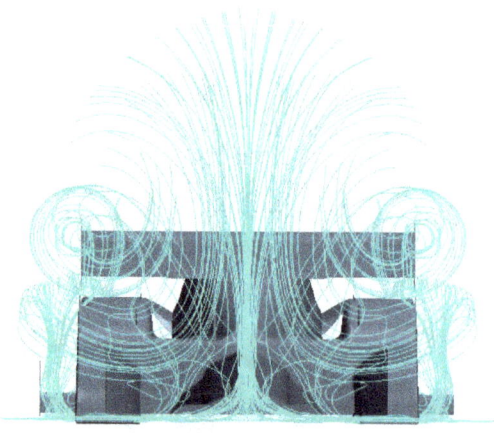

Other elements can also be considered as front wings. In the case of LMP1 Le Mans series the bodywork acts as a front wing:

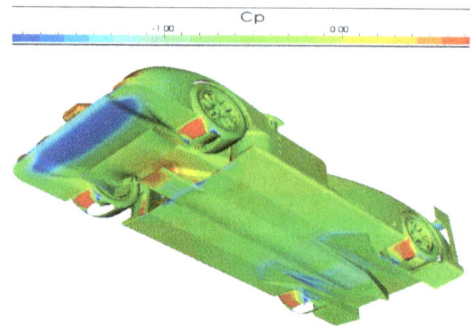

FRONT WING FOR ACCELERATION

Let's take a look at a special arrangement of the upper front wing that used McLaren for many years to boost power output: It was a kind of arch going over the nose and anchored on both sides of the front wing.

Renault took a different design alternative, but with the same goal:

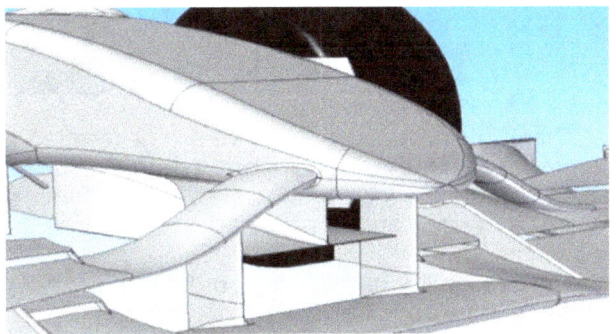

Ferrari and other teams also installed something similar, some with supports to prevent excessive deflection at the center:

This was the 2008 Ferrari:

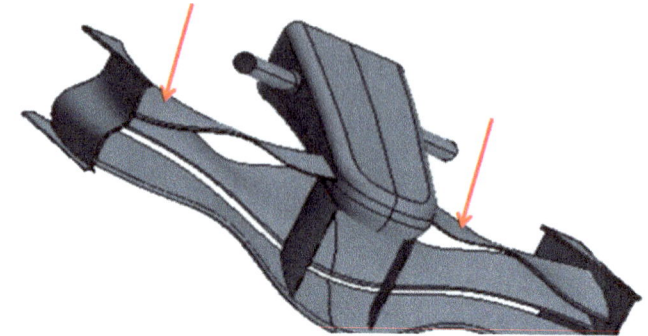

       Thank to these extra wing we are generating even more downforce in the central part. This makes it easier to reach the necessary downforce so lateral flaps don't need an extreme angle of attack that could affect sidepod inlets.

       The availability of this wing on the top also responds to a dynamic issue:

       When a car is cornering or braking, the wing's distance to the ground is very small; When accelerating, the car transfers mass to the rear and raises the nose of the car; The front spoiler is designed to work optimally near the asphalt; when accelerating at the end of a corner the car can understeer excessively , unless there is "something" that produces downforce at the front; this device is this top wing: The wing is not affected by the distance to the ground and works perfectly at all times regardless of speed or performance that the car has.

STRAKES

       They consist of elements that are placed in the bottom of the front spoiler and play several important functions:

It's funny, but this system increases lift on the rear axle, rather than on the front:

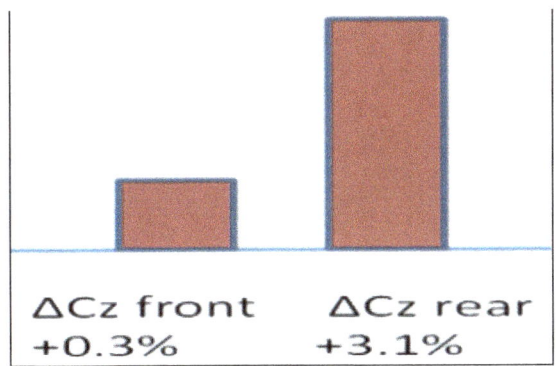

But it also increases drag:

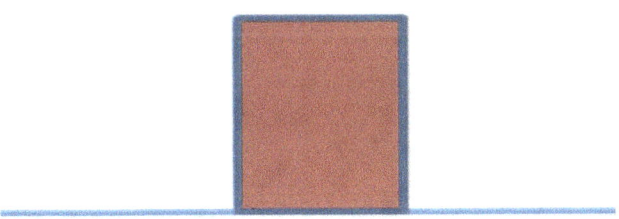

Cx rear +1.3%

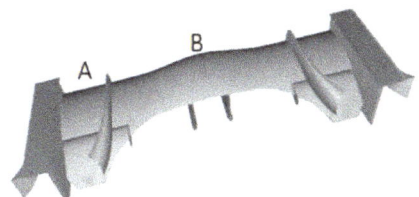

In the area "A", we are increasing speed; Area "B" is increasing the depression and therefore downforce is being generated.
Front view:

→ There is one function very important of front wing; that is vortex generating (may be, this point, would be placed in Ground and Diffuser Chapter, but here, we think is correct also).

The front wing, is the first device aerodynamic in a car; behind it, there are:
- Sidepod.
- Ground / Diffuser.
- Wheels.
- Etc….

For that, is possible to generate from front wing, vortices in order to do some think; for example:

These vortex for example, are created to help the ground effect; with these vortices, it will be more difficult escape through the sides of the floor.

These vortices can produce air turbulent, which is good for heat transfer in sidepod.

Is very important not forget that these vortexes are produced by depression tube; this tube produce a suction of another air mass.

And finally in:

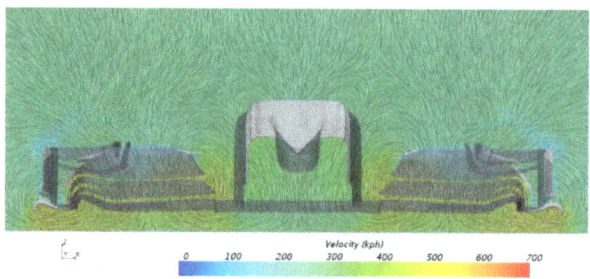

Obviously, these vortex can exist for producing stability and also seal the ground or adapt the air flow to behind attached to sidepods.

Vortex Generators:

VGs are known as the first fix for flow separation in aerospace applications. The idea is to redistribute momentum by bringing high momentum flow outside the boundary layer to the inside. For this purpose, VGs are typically of the height of the local boundary layer and placed about 20 times local boundary layer height ahead of the separation point. The boundary layer scale VGs are used in the automotive applications as well to prevent flow separation.

However other than the traditional boundary layer VG, large scale VGs can often be seen on racing cars. These VGs interact with the mean flow, instead of the viscous boundary layer. They can be used to generate vortex suction (add downforce) and help turning/guiding the flow. Various devices, such as barge boards, turning vanes, front wing strake, etc., can be possibly generalized into large VG category (writer's point of view).

Get now, one front wing:

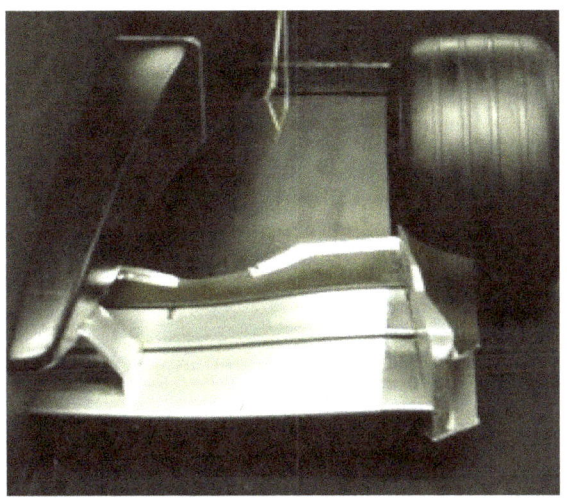

Analyzing the flow into wing:

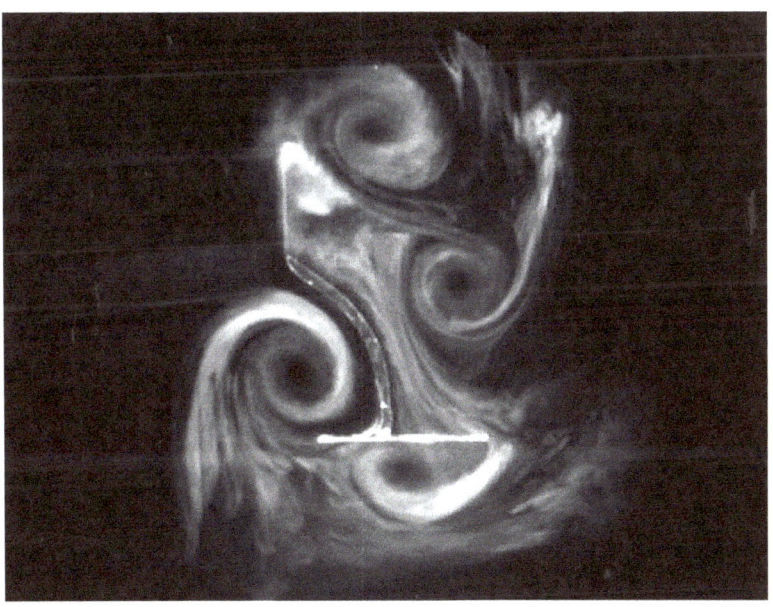

Amazings vortex
¡¡

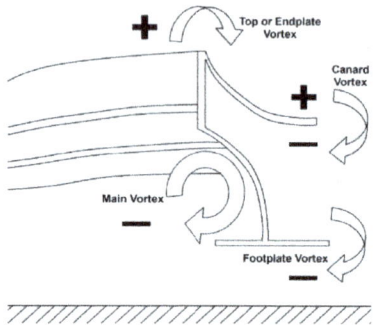

What is the behavior of these vortex, trough front wheel ?

There is a deflection (interaction in general):

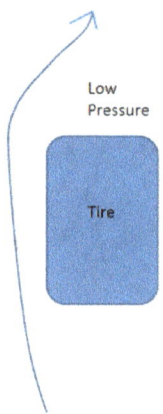

There is a value very important for defining a vortex; that is named Vorticity; for calculating that:
1.

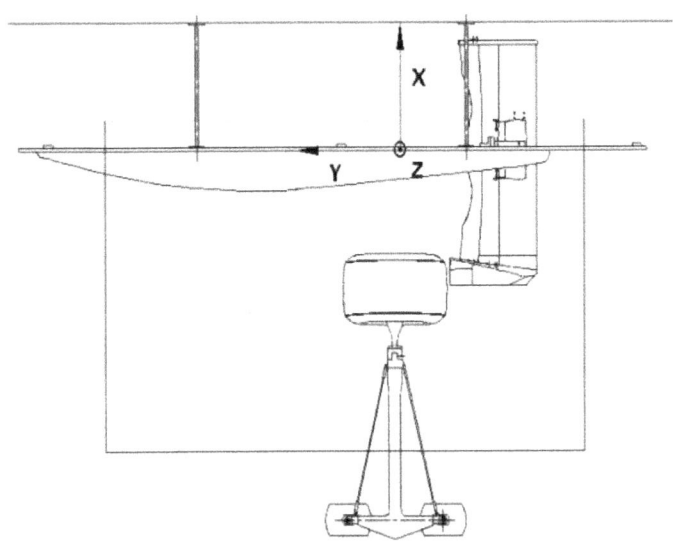

$$\left(\omega_y\right)_{i,j} = \frac{\Gamma_{i,j}}{4\Delta X \Delta Z}$$

$$
\begin{aligned}
\Gamma_{i,j} = \ & \tfrac{1}{2}\Delta X(U_{i-1,j-1} + 2U_{i,j-1} + U_{i+1,j-1}) \\
+ \ & \tfrac{1}{2}\Delta Z(W_{i+1,j-1} + 2W_{i+1,j} + W_{i+1,j+1}) \\
- \ & \tfrac{1}{2}\Delta X(U_{i+1,j+1} + 2U_{i,j+1} + U_{i-1,j+1}) \\
- \ & \tfrac{1}{2}\Delta Z(W_{i-1,j+1} + 2W_{i-1,j} + W_{i-1,j-1})
\end{aligned}
$$

This last value of Circulation, is function of U and W, velocity in the two axis X and Z.

2.

Another method for calculating the vorticity, is based in displacement points in a area:

$$\Gamma_1(P) = \frac{1}{N}\sum_S \frac{(PM \wedge U_M)\cdot z}{\|PM\|\cdot\|U_M\|}$$

$$\Gamma_2(P) = \frac{1}{N}\sum_S \frac{\left(PM \wedge (U_M - \overline{U}_P)\right)\cdot z}{\|PM\|\cdot\|U_M - \overline{U}_P\|}$$

Center vortex:

Where P is a fixed point in the measurement domain. S is a two-dimensional area surrounding P, referred to as the window size, whose size must be defined by the user. N is the number of data points (M) that are located within S, and UM is the velocity vector at point M.

Finally, z is a unit vector normal to the plane of measurement, and Γ1 (or Γ2) is calculated for all points in the measurement domain and then used to determine the center of the vortex. For an ideal axisymmetric vortex, the maximum value of |Γ1| is 1. For each possible center position, the Γ1 criterion calculates the degree to which the flow rotates around this point. In this work, near the vortex core, |Γ1| was found to reach values between 0.9 and 1.

The only difference between the Γ1 and Γ2 criteria is that the average velocity over the window, away. This takes into account any uniform flow within the plane of rotation. With this method, when |Γ2| is greater than approximately 2/π, P is assumed to represent a point in the vortex.

Finally, another method:

3.

The mathematical model for the flow velocity in the circumferential θ-direction in the Lamb–Oseen vortex is:

$$V_\theta(r,t) = \frac{\Gamma}{2\pi r}\left(1 - \exp\left(-\frac{r^2}{r_c^2(t)}\right)\right),$$

with

- r = radius
- $r_c(t) = \sqrt{4\nu t}$ = core radius of vortex.
- ν = viscosity, and
- Γ = circulation contained in the vortex.

The radial velocity is equal to zero.

The associated vorticity distribution[13] in the vortex-filament-direction (here $\hat{z}$) can be found with the curl

$$\omega_z(r,t) = \frac{\Gamma}{\pi r_c(t)^2}\exp\left(-\frac{r^2}{r_c^2(t)}\right),$$

An alternative definition is to use the peak tangential velocity of the vortex rather than the total circulation

$$V_\theta(r) = V_{\theta\,max}\left(1 + \frac{1}{2\alpha}\right)\frac{r_{max}}{r}\left[1 - \exp\left(-\alpha\frac{r^2}{r_{max}^2}\right)\right],$$

where $r_{max}(t) = \sqrt{\alpha r_c(t)}$ is the radius at which v_{max} is attained, and the number $a = 1.25643$, see Devenport et al.

The pressure field simply ensures the vortex rotates in the circumferential direction, providing the centripetal force

$$\frac{\partial p}{\partial r} = \rho\frac{v^2}{r},$$

Is very important to know where the vortex is, that is:

- Vortex center.
- Circulation value.

But is more important may be, the interaction between vortexes.

If we have two vortices:

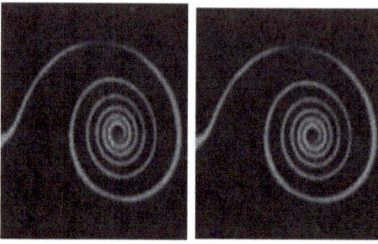

And we have two vortices:

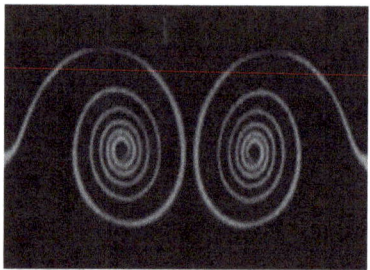

The dynamic full (together), not will be the same; that is very important.

For example: we have two vortices with the same direction rotation:

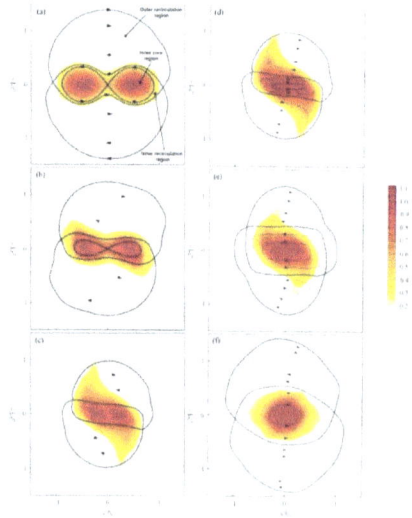

As the merging process continues (d) and (e), the inner region expands and recaptures some of the anti-symmetric vorticity of the filaments, which reduces this induced velocity. Hence the vortex diffusion mentioned earlier is required to finally bring the separation distance to zero. The process concludes with the symmetric vortex seen in (f).

These unions or alteration, depending of intensity (vorticity) of vortex; if there is one vortex, bigger than other vortex (red and green), can produce that: (each line correspond one context or size):

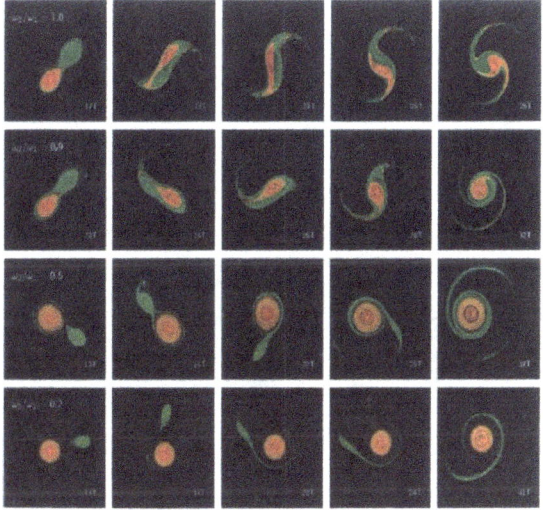

Is very pretty and amazing, this evolution and the geometry created.

But we know already, one vortex can produce another vortex, from track:

Contour plot of vorticity (s downstream. The opposite signed vorticity from the boundary layer has been convected away from the wall by the vortex. (b) schematic diagram explaining the origins of the rebound effect and of the stream-wise boundary layer.

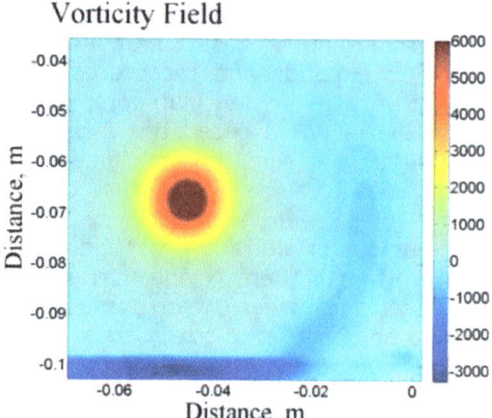

The whole system is convected in the direction of the green arrow. Another factor which contributes to the increase in b in the C1 configuration downstream is simply the increase in size of the streamwise boundary layer:

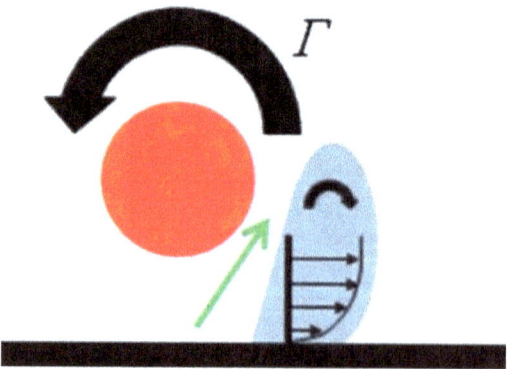

We can see, that this theme of vortex, creation, evolution, interaction and finish, is very complicate, but necessary. I´m sorry….

→ **The vortex is very important; may be the most important in race car aerodynamic; these, allow "play with the air", in order to help to go where is needed iiii to generate vortex, is to wrap air….**

Note important:

Let a car with a sidepod; we need transport air, to rear, above diffuser:

We can place in sidepod, a deflectors:

If these deflectors not produce vortex, the air "has little energy"; so is practically impossible that the air reaches the area:

But, is we rotating this air, as a vortex, is easier transport it:

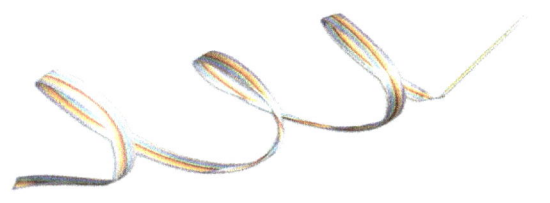

In fact, from a point out of profile-wing, a vortex is created:

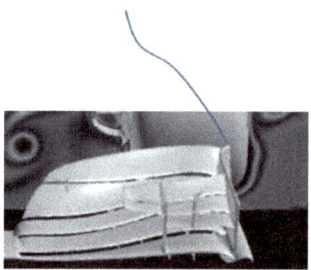

This "way" is as a snake (flying....); this snake is a depression tube:

More thinks about:
What should be the direction of rotation ? think about football:

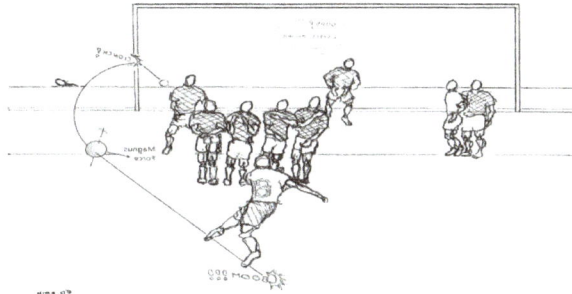

The answer: the first image is the correct:

So the deflectors or vortex generators, should be these:

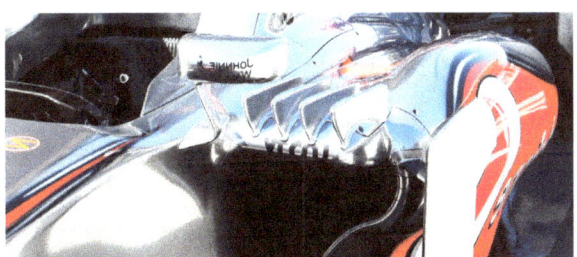

→ *Question important:*

We have a delta wing flat (without profile) with 0º of incidence angle; are produced vortex ?

     Yes yes. Because the air wants to fill the depression.
     What depression ?
     The depression produced by the friction of air (reduction speed) with the surface:

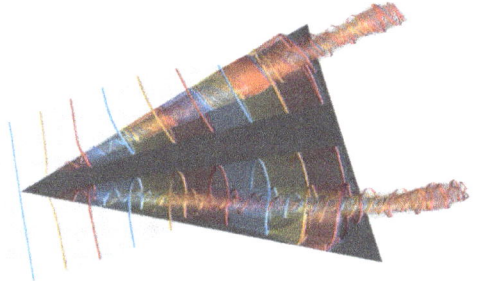

But…. If up and down of surface, have the same friction, what produce the difference of pressure ? is possible so, change the friction….

That is the same than circle and the formation so, of Karman vortex: the circle is symmetric…. There are small variations of pressure….:

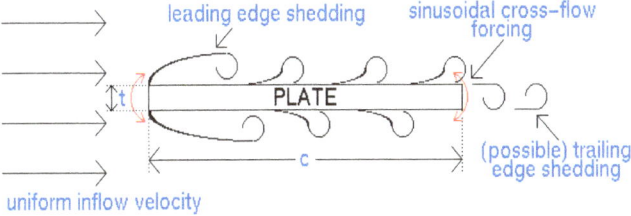

For example, in the next car, are produced vortex ? yes, by the same:

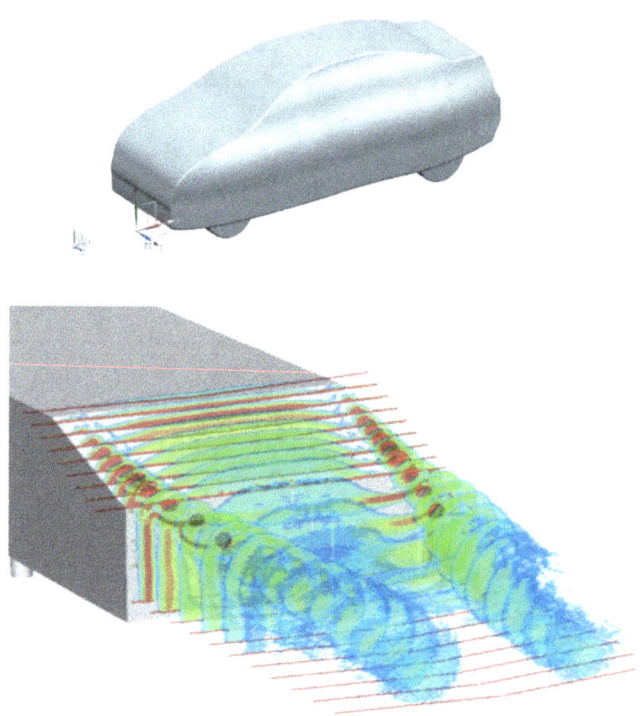

These, we know, are named Pilar Vortex. These vortex are important in order to improve the stability.

➜ If the quantity of vortex is big, the stability is big, but the drag is also big. That is the same that the grip between tires and track: if there is a lot grip, there is a lot drag….

Summary:
Is necessary so, control the downforce of front wing to specific value or below, rather than maximizing it.

This monograph explores the physics of an eddy formed between two layers moving at different speeds in a flow field. High speed flow V1 (Depicted below) moves over a lower speed portion moving at V2.

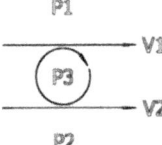

In between the two linear parts of the flow, a third zone forms: where the flow is undergoing a rotational motion, superimposed on linear motion in the same direction as the other zones. All three zones, the two linear portions and the eddy, have the same static pressure so the pressure P1, P2 and P3 are the pressure depression due to motion, the dynamic head. Since the total energy (sum of kinetic and potential, or pressure, energy) is a constant for all of the flow zones (heat and work are not being added or removed) and if we ignore frictional heating, it is possible to calculate the unknown pressures and velocities using Bernoulli's equation. Understanding the relationship between rotational motion and pressure forces is required to do this. Consider the figure below:

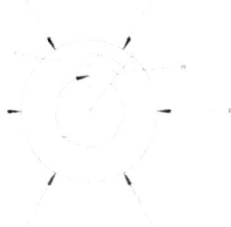

Pressure forces (the external inward pointing arrows) act on the perimeter eddy. Opposing the pressure force is the force required to contain the outward acceleration or centrifugal force (Note: Centripetal: Latin for towards the center and Centrifugal is Latin for traveling away from center) of the rotating flow. Outward acceleration is $\omega^2 r$ time the mass of the vortex. Inward pressure forces equal and oppose the outward rotational acceleration when balance is achieved. Thus from force equals mass times acceleration:

$$PA = \frac{m\omega^2 r}{g}$$

Assigning a depth of "d" to the above figure and computing the mass of fluid using density in the vortex by performing the following integral along with an expression of the outer area yields:

$$P\pi 2rd = \left(\rho d \int_0^r 2\pi r dr \right) \frac{\omega^2 r}{g}$$

Simplifying:

$$P = \frac{\rho \omega^2}{g} \int_0^r r dr = \frac{\rho \omega^2}{g} \frac{r^2}{2} + c$$

The "c" in the above equation is the pressure at the center or r =0. With a relationship between pressure, rotational / radial velocity (omega = V_rr) and radius it is possible to explore what happens at the steady state velocity interface. Writing the Bernoulli form of the energy equation:

$$\frac{V_1^2}{2g} + \frac{P_1}{\rho} = \frac{V_2^2}{2g} + \frac{P_2}{\rho} = \frac{V_3^2}{2g} + \frac{P_3}{\rho} + \rho \omega^2 \frac{r^2}{2g}$$

Since P_3 is equal to:

$$P_1 - P_2 = P_3$$

Gathering the velocity and pressure terms together:

$$\frac{V_1^2}{2g} - \frac{V_2^2}{2g} + \frac{P_1}{\rho} - \frac{P_2}{\rho} - \frac{P_3}{\rho} = \frac{V_3^2}{2g} + \rho \omega^2 \frac{r^2}{2g\rho}$$

Substituting for P_3 for $P_1 - P_2$ yields:

$$\frac{V_2^2}{2g} - \frac{V_1^2}{2g} = \frac{V_3^2}{2g} + \frac{\omega^2 r^2}{2g}$$

Multiplying both sides of the equation by 2g yields:

$$V_2^2 - V_1^2 = V_3^2 + \omega^2 r^2$$

Assigning values to V1, V2 and V3 still leaves two unknown variables ω and r. Let us assume that the radial velocity (ωr) of the spinning fluid is one half of the velocity difference ($V_1 - V_2$) 2. This assumption is based on the fact, that the dynamic pressure forces at the top and bottom or the rotating fluid are equal and opposite and thus in equilibrium. Further, for simplicity, it is assumed that V_3, the vortex's linear velocity, is zero. Thus: $V_3 = 0$, ωr = (V_2-V_1)/2, V_1=2, V_2=1

Thus:

$$\omega r = \frac{V_2 - V_1}{2} = \frac{1 - 2}{2} = -.5$$

and:

$$r = \frac{-.5}{\omega}$$

Substituting the above into the equation for ωr:

$$1^2 - 2^2 = \omega^2 \frac{-.5^2}{\omega}$$

Solving for ω:

$$\omega = \frac{V_2^{\,2} - V_1^{\,2}}{\left(\frac{V_2 - V_1}{2}\right)^2} = \frac{1^2 - 2^2}{-.5^2} = \frac{1 - 4}{.25}$$

$$\omega = -12$$

With ω, r is:

$$r = \frac{-.5}{\omega} = \frac{-.5}{-12} = 0.04167$$

The Charts below show omega (ω) and r when V_1 is 1, 10, 100 and 1000 and where V_2 varies from 0 to each value of V_1:

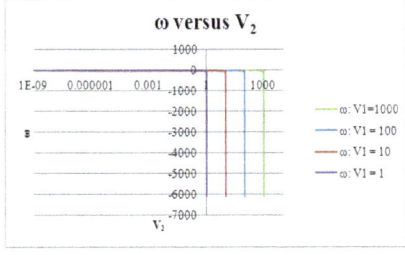

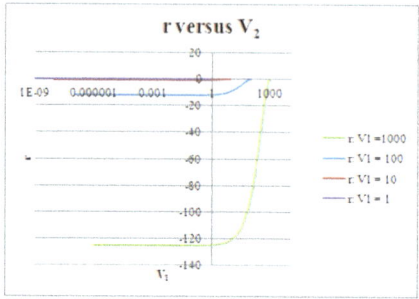

r versus V_2

Legend:
- r V1 =1000
- r V1 = 100
- r V1 = 10
- r V1 = 1

At equilibrium ω varies from 4 to 6076 as the velocity gradient decreases and r varies from r = V_1 times 0.125 to zero as the velocity gradient decreases High velocity flows have the potential to develop quite large eddies. At a boundary, a flow field develops many small rotating layers and as the flow moves further along the balance will be lost when the flow field changes due to momentum   pressure lose, known as diffusion, to other portions of the flow field, momentum loss to frictional losses etc.   At some point the balance will be broken and the rotating momentum will become linear momentum at the point where the balance loss occurs. This will accelerate the surrounding flow in the direction perpendicular to where the imbalance occurs. A second eddy could form or perhaps curvilinear motion. Clearly this process would appear to be chaotic.

The next monograph will find how long it takes to form an eddy.

REAR WING

This element is attached to the rear of the car, its main function is to generate downforce; unlike other aerodynamic parts of the car, the rear spoiler does not have the function of redirecting flow to a specific area of the car as there are no elements behind it; it is responsible for generating downforce through 2 methods:

- Generating downforce on its own.
- Generating downforce indirectly, optimizing the diffuser and therefore the

car's floor (ground effect).

In elements installed where the ground effect is not important, is possible to place these elements in order to improve the downforce or efficiency of rear wing; in general to rear car:

This concept is essential for allowing a good "full" design.

Consider the two functions:
The rear wing consists of one or more parts (wings); these wings generate downforce.

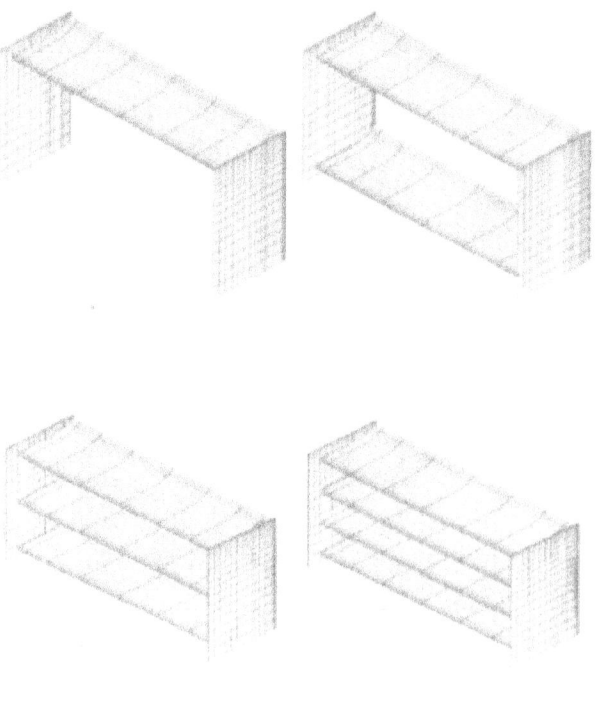

The rear wing also includes other elements whose functions are:

- To improve the operation and purpose of the wings.
- To provide a base for placing other aerodynamic elements.
- To attach the wing to the chassis.

As we saw previously with the front wing, the rear wing originally was also that: a wing creating downforce exclusively:

But designs have evolved under the three functions outlined above and depending on which category the car races:

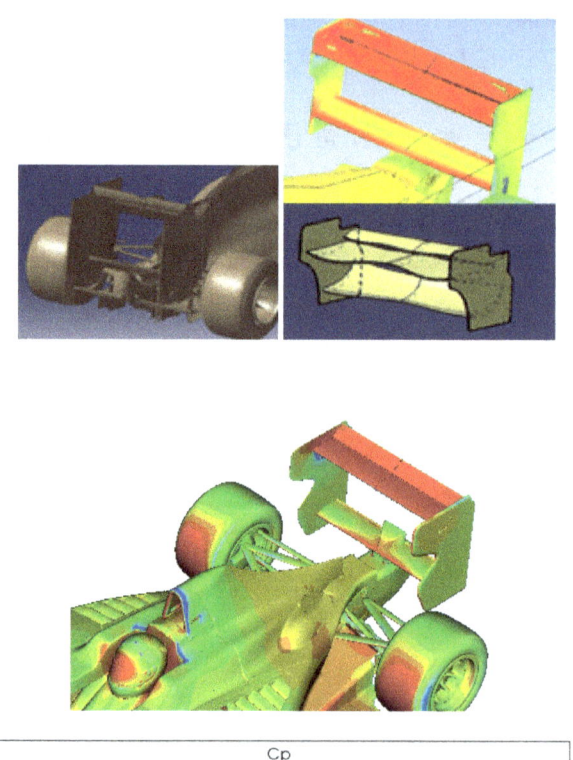

	Cp		
-1.50	-1.00 -0.50	0.00	0.50

The rear wing, also produce vortex:

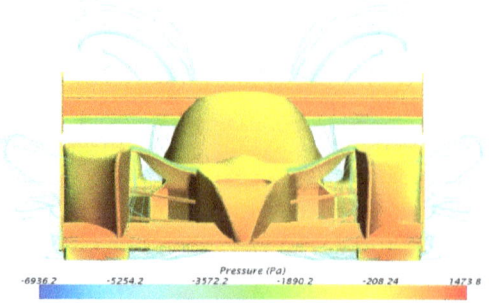

Pressure (Pa)

-6936.2   -5254.2   -3572.2   -1890.2   -208.24   1473.8

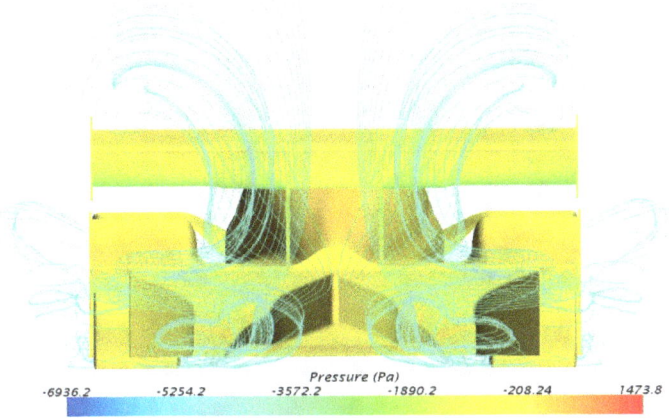

Pressure (Pa)

-6936.2   -5254.2   -3572.2   -1890.2   -208.24   1473.8

In order to maintain the gap between different flaps or parts in rear wing, there is a little piece:

This device produce a turbulence in rear wing surface. For reduce this turbulence, there is a "crack" over the wing:

Similarly to front dive planes you can see these devices at the rear of some racing cars:

One of the important things to consider is the position of the rear wing; if we are analyzing GT type cars, we can follow these simple rules of design: depending on the size and position we obtain variations of downforce and drag:

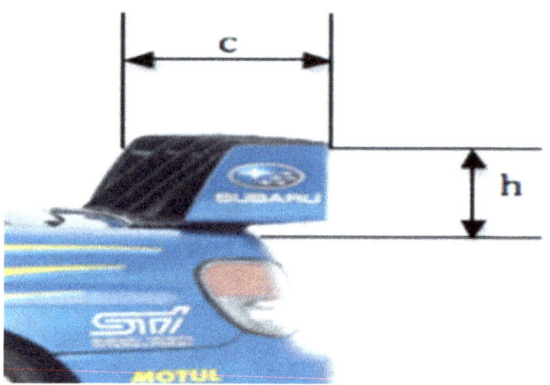

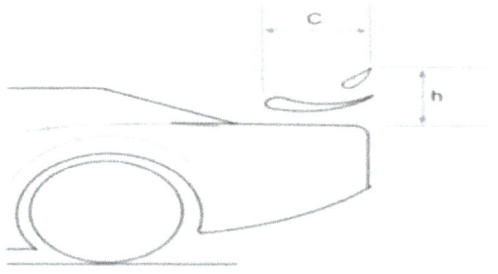

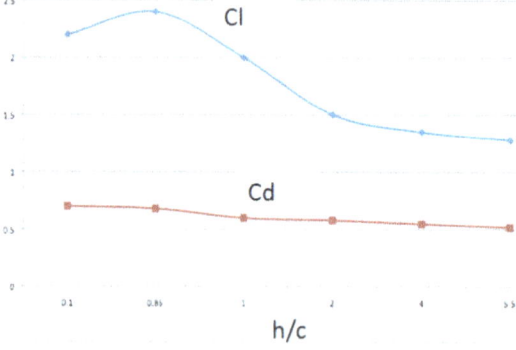

Another important factor in the design of a rear wing is the generation of turbulence at the back; to generate downforce, wings send up air; this flow is turbulent ruining the good aerodynamic behaviour of a pursuing car; FIA decided in favour of the show and increasing the amount of overtakes per race, to mitigate the turbulence re-designing the rear wing of a F1; the concept is the following:

This design did not generate turbulence in the central part of the car but at the extremes; this meant cleaner air for the pursuing cars.

Another objective of a rear spoiler is to help the diffuser obtain its maximum potential; as discussed below, the floor of the car and the diffuser are two of the most important elements in terms of the amount of downforce generated, so increasing their performance is necessary.

A rear wing normally comprises of two endplates and a number of wings at the top part; many wing-sets include another wing at the bottom called diffuser wing. This wing is the one responsible for assisting the diffuser to work as it should; this wing on its own produces little downforce, but indirectly it is generating a lot by making the diffuser function correctly:

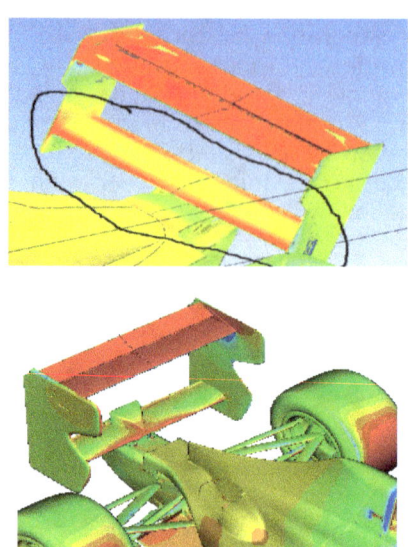

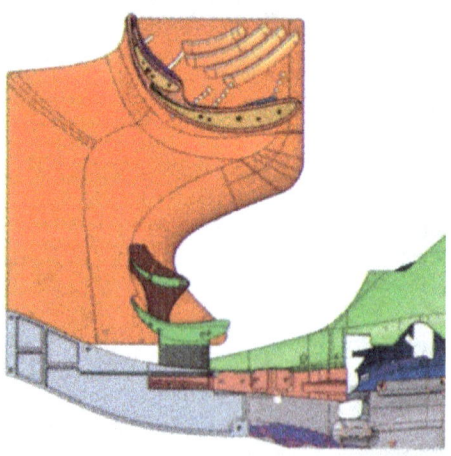

Exist examples "rares", in order to adapt the behavior to corner, if there are, for example, more corner in right than left:

Examples in CAD for wings Race Cars:

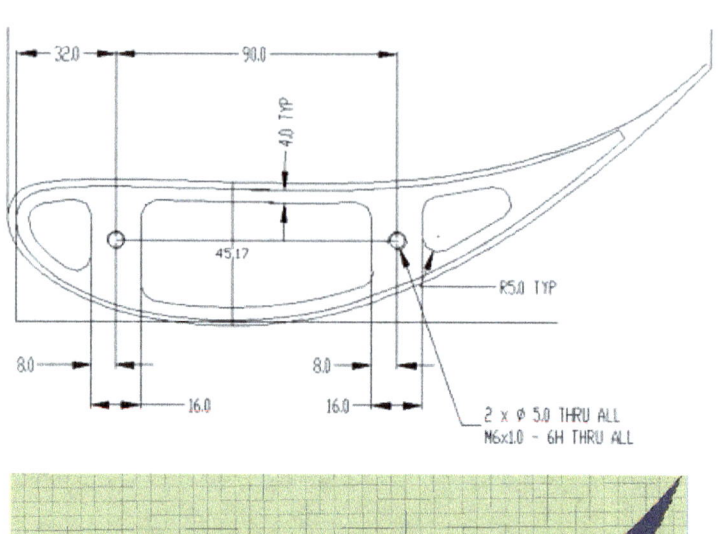

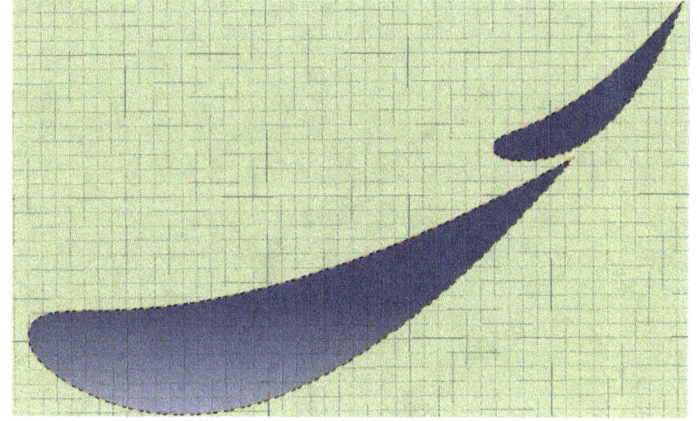

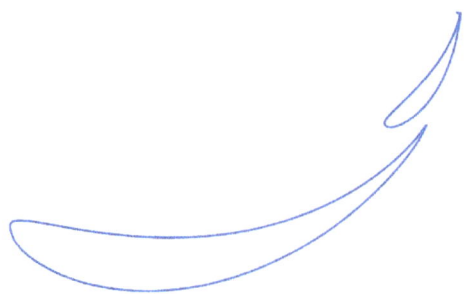

Be 153-055

Ro=2.38 Xo=2.38 -Yo=0.84

x%	0	1.25	2.5	5	7.5	10	15	20	30	40	50	60	70	80	90	95	100
$-Y_v$	.84	-1.34	-2	-2.71	-3.1	-3.32	-3.52	-3.52	-3.15	-2.42	-1.51	-.44	.73	1.67	1.7	1.15	.1
$-Y_o$	.84	3.45	4.61	6.26	7.48	8.45	9.88	10.84	11.86	12.04	11.5	10.22	8.31	5.89	3.13	1.66	.2

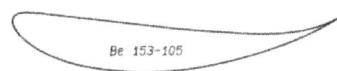

Be 153-105

Ro=2.12 Xo=2.12 -Yo=0.86

x%	0	1.25	2.5	5	7.5	10	15	20	30	40	50	60	70	80	90	95	100
$-Y_v$	.86	-1.06	-1.45	-1.59	-1.44	-1.18	-.59	.01	1.18	2.39	3.49	4.36	5.03	5.19	3.97	2.48	.1
$-Y_o$	.86	4.09	5.61	7.86	9.57	10.95	13.05	14.52	16.25	16.87	16.5	15.07	12.72	9.6	5.58	3.09	.2

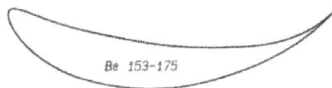

Be 153-175

Ro=1.77 Xo=1.77 -Yo=1.62

x%	0	1.25	2.5	5	7.5	10	15	20	30	40	50	60	70	80	90	95	100
$-Y_v$	1.62	.04	.07	.78	1.72	2.68	4.42	5.87	8.16	10.03	11.33	11.86	11.71	10.64	7.46	4.54	.1
$-Y_o$	1.62	5.71	7.75	10.86	13.29	15.28	18.35	20.56	23.29	24.54	24.36	22.65	19.6	15.36	9.36	5.31	.3

Be 122-125

Ro=2.85 Xo=2.85 -Yo=3.59

x%	0	1.25	2.5	5	7.5	10	15	20	30	40	50	60	70	80	90	95	100
$-Y_v$	3.59	1.22	.57	.06	.02	.22	1.01	2.1	4.5	6.62	8	8.44	7.82	6.15	3.5	1.84	0
$-Y_D$	3.59	6.35	7.54	9.24	10.51	11.54	13.14	14.27	15.54	15.73	15.06	13.52	11.23	8.2	4.45	2.3	.1

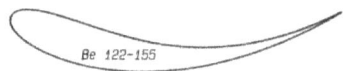

Be 122-155

Ro=2.77 Xo=2.77 -Yo=3.22

x%	0	1.25	2.5	5	7.5	10	15	20	30	40	50	60	70	80	90	95	100
$-Y_v$	3.22	.9	.34	.03	.19	.59	1.79	3.25	6.29	8.82	10.39	10.77	9.87	7.72	4.39	2.33	0
$-Y_D$	3.22	6.18	7.51	9.44	10.93	12.17	14.12	15.58	17.37	17.93	17.39	15.83	13.29	9.79	5.37	2.81	.1

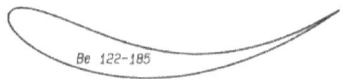

Be 122-185

Ro=2.71 Xo=2.71 -Yo=2.94

x%	0	1.25	2.5	5	7.5	10	15	20	30	40	50	60	70	80	90	95	100
$-Y_v$	2.94	.67	1.9	.07	.43	1.03	2.63	4.47	8.13	11.07	12.8	12.09	11.9	9.25	5.25	2.79	0
$-Y_D$	2.94	6.11	7.57	9.75	11.46	12.89	15.22	16.99	19.28	20.16	19.75	18.13	15.31	11.34	6.25	3.28	.1

This profiles set, is good for low speed or low incidence angle (lower than 12 degrees) but not for others angles. We will see that, in other book-volume, and also the endplate design and his improvement.

About the importance of gap between main wing and flap, we can see the next results:

Study the Influence of a Gap between the Wing and Slotted Flap over the Aerodynamic Characteristics of Ultra-Light Aircraft Wing Airfoil.

Cvetelina Velkova1 , Michael Todorov1 and Guillaume Durand2 1 Department of Mechanics, Faculty of Transport, Technical University, Sofia 1000, Bulgaria 2. ECAM, Strasbourg-Europe, 2 rue de Madrid - 67300 Schiltigheim, France.

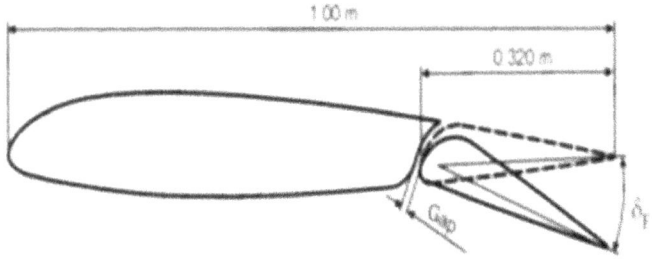

NACA 23012 airfoil with single slotted flap

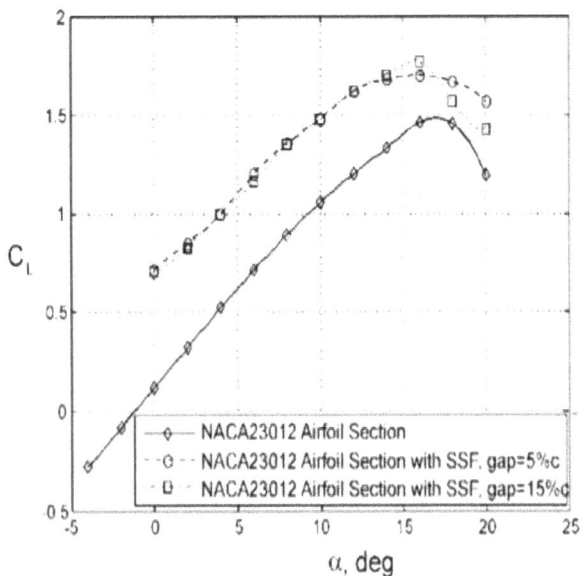

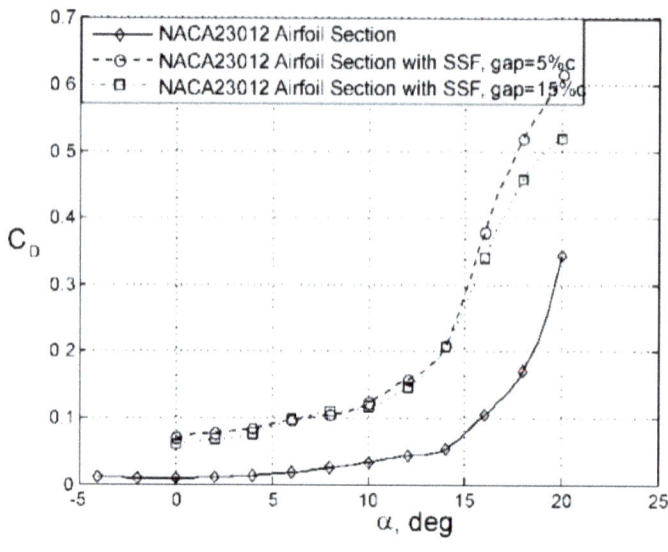

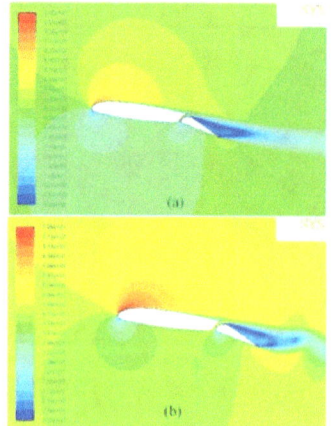

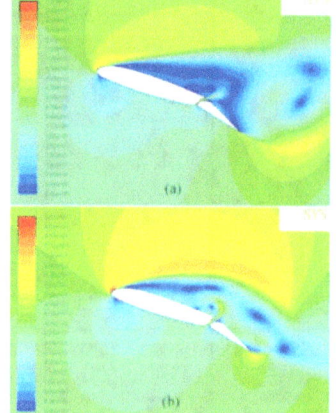

Velocity field around NACA 23012
airfoil with a single slotted flap at α=6°, (a)
gap size 0.005 m, and (b) gap size 0.015 m

Velocity field around NACA 23012
airfoil with a single slotted flap at α=16°, (a)
gap size 0.005 m, and (b) gap size 0.015 m

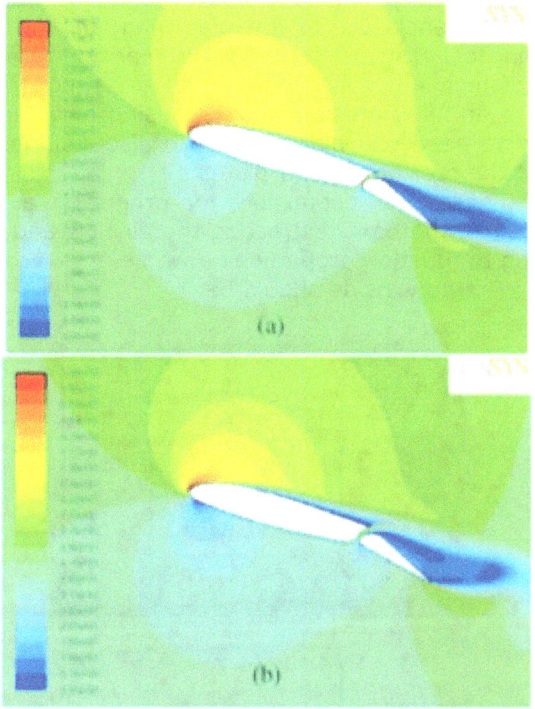

Velocity field around NACA 23012
airfoil with a single slotted flap at α=14°, (a).
gap size 0.005 m, and (b): gap size 0.015 m

We know that if we modify the angle of attack of the rear wing the downforce of the front wing and the global center pressure is affected; seems "illogical" but true; think a case where you change the rear wing, the diffuser is directly affected; this in turn produces a change in the direction of flow going underneath the car producing a measurable change in the aerodynamic forces; It is a type of chain reaction from the back to the front.

When we cover in depth the diffuser and the ground effect, we will take a closer look at this wing.

The end-plates can be used to extract air flowing through the enclosure of the rear wing; the air incident to a wing generates a series of turbulences between the areas of high and low pressure; at the wing ends high pressure and low pressure meet causing turbulence that is why we need end-plates; We can also include Gurney flaps at the rear of the end-plates:

These are responsible for producing a vacuum inside the "wing box", extracting the air flow and causing the wing to function properly as a whole.

The flow inside should be:

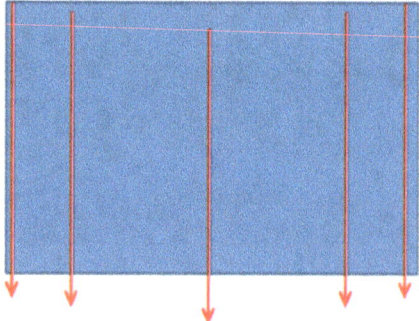

But in reality, the flow on the outside is not the "ideal"

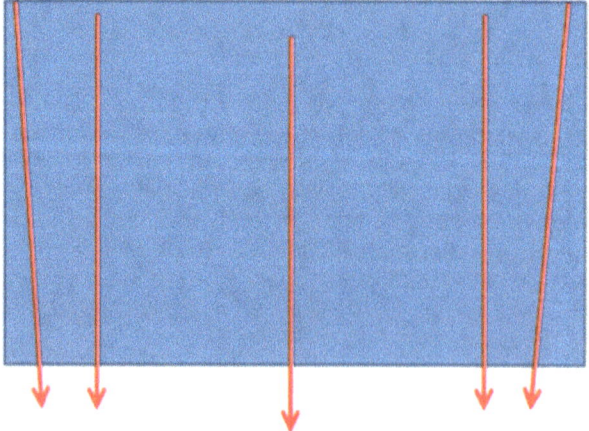

That Gurney flap, is installed vertically also in the end-plates:

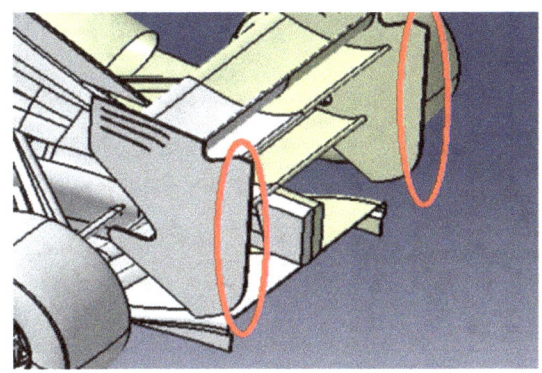

There is another method to achieve the same thing by installing a flake; we'll see it later
Note:
Geometrically, the rear wing of a F1 is a box, in which the wing and diffuser is located:

The rear wing "box", tries to create a channel through which air circulates inside it; we all know how aircrafts powered by ducted fans work:

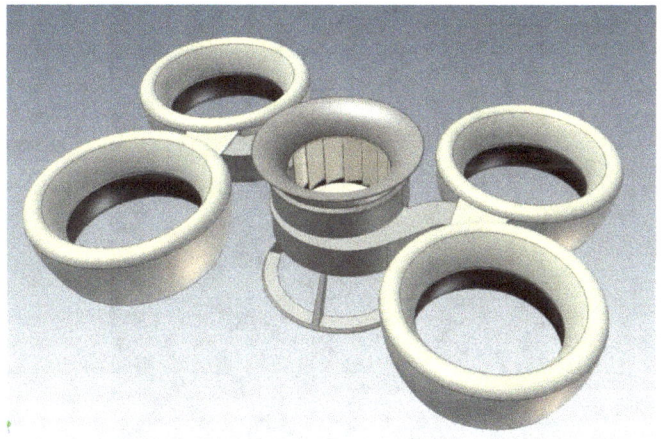

The essence of this system is to optimize and improve the performance of the propeller; in our case, the objective is to optimize the wing's functioning. The "box" wing works as multiple venturi tubes which are responsible for accelerating the airflow.

Imagine that you are forcing a ball to levitate with a hairdryer; for this, the ball will levitate at a certain height, depending on the velocity of the dryer air:

Suppose now that while the ball "levitates" we put a tube so that the ball is entirely within the tube:

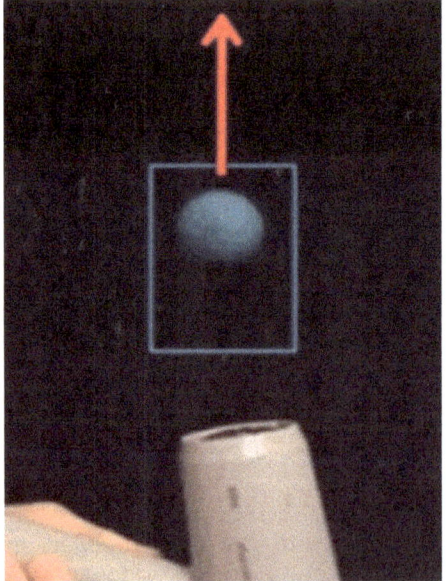

We will see that the ball is pulled up; this fact is because inside the tube a considerable increase in speed occurs, pushing up the ball; we can obtain the same effect in a rear wing: increasing the air speed below the wing reducing the pressure and increasing the downforce generated.

The notch that almost all rear wings have in the upper back is responsible for reducing the turbulence thus reducing drag:

Here we can perfectly see the concept of a "wing box":

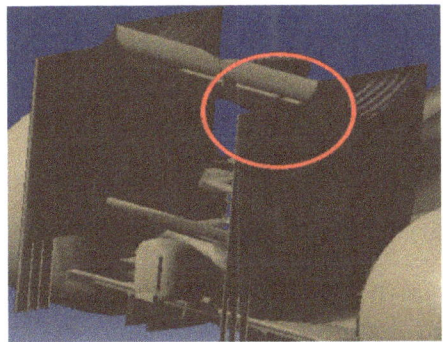

Also the lovers in side (filling the rear vortex): that, is the same than the nut in wheels in order to fill the rear wheel depression):

Sample:

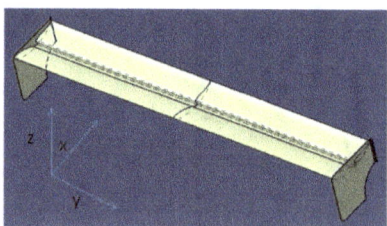

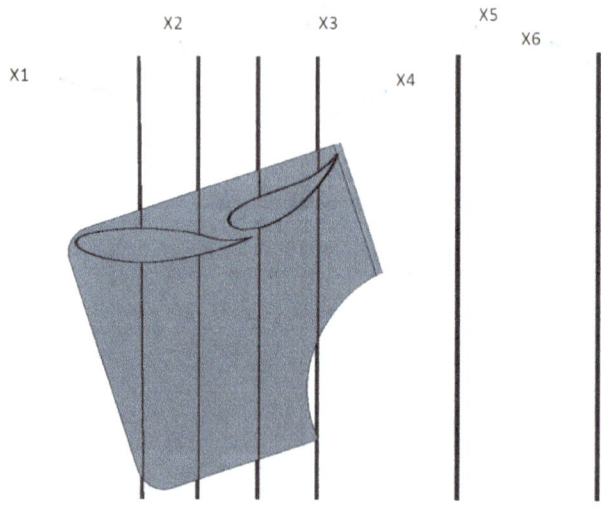

PRESIÓN – SECCIÓN X5

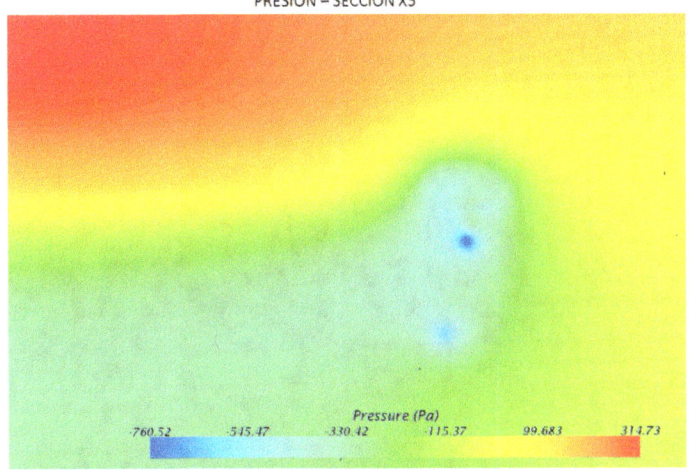

PRESIÓN – SECCIÓN X6

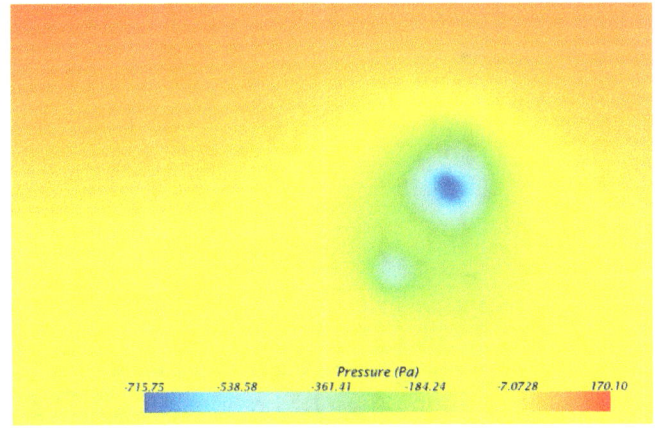

VELOCIDAD – SECCIÓN X5

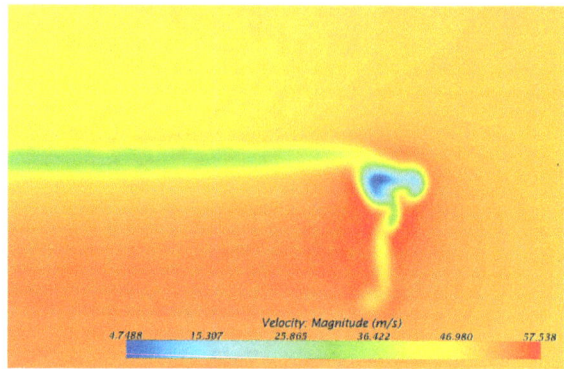

VELOCIDAD – SECCIÓN X6

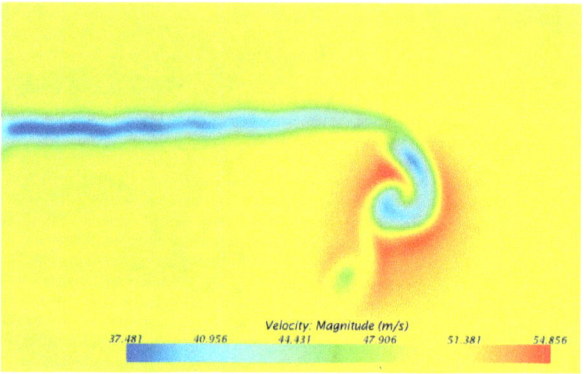

One of the important things that should be taken in to account is to find a way to reduce the global drag using the rear wing; we can place the wing's flap with a null angle of attack keeping the centre of pressure at the same point (This happens in the Monza Circuit, for example).

➜ For example, Mc Laren – 2016:

This device allow to introduce more air flow in a "cave" rear aileron. That produce with the same incidence angle, more downforce, so if the angle is lees the drag also is less. Amazing concept iiii The same concept, we can see it in the barge boards or some thinks as that:

AERODYNAMIC DEVICES OR SMALL WINGS

One of the many problems that an aerodynamicist will find, is designing wings or small aerodynamic appendages.

We have already seen how you can design and what features must all wings include; but when the size is small, everything changes.

Previously we have seen how incorporating end-plates (increased "effective" area) affected the characteristics of a wing; we will look at the same but applied to small wings.

We define the aspect ratio as AR = SPAN / CHORD

We define "effective AR" as the "real" extension of a wing; this is not the real dimension of the wing but the effective dimensions due to installation of endplates.

We define "bv" as the height of the end plate, "cv" as the rope screen tip and "cH" the wing chord; this way, we can calculate the effective length of the wing:

$$AR_{eff} = AR\left(1 + \frac{1}{2}\frac{bv}{cv}\left(\frac{cv}{cH}\right)^2\right)$$

We can use the following expression for the same calculation; however, it does not take into account the installation of the plate tip:

$$A\operatorname{Re} ff = AR\left[1 + 1.9\left(\frac{h}{b}\right)\right]$$

And finally, we can calculate the wing's lift coefficient, using:

$$Cl = \pi \cdot (A\operatorname{Re} ff) \cdot \tan(incidence - angle)$$

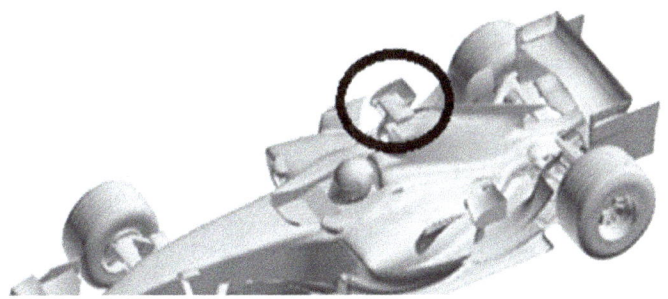

Remember something important:

Incorporating end-plates are necessary in small aerodynamic parts to ensure that the whole area of the device generates downforce avoiding the air to escape sideways.

If the goal is create vortex, is not necessary this end plates; as example, in sidepod leading edge flickups (devices in barge board).

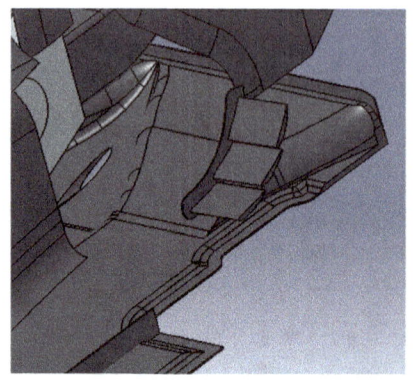

The next wing, need endplate, if the main goal is "only" adapt the flow in to sidepod (is possible to generate vortex....):

One of these small devices nowadays popular in F1 is the "Monkey Seat"; it is located strategically to generate downforce on it own, apart from helping the diffuser.

The next "Wings" as a Monkey Seat, reduce the drag in rear depression, and so, the full drag:

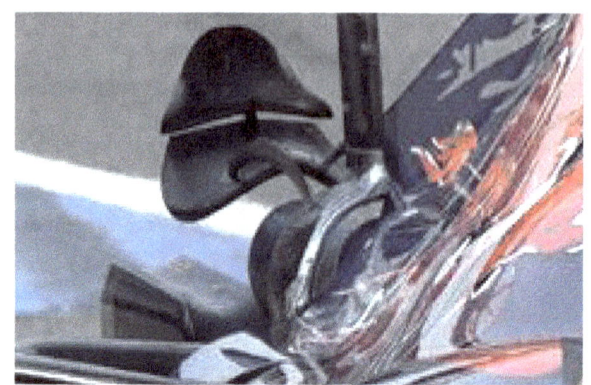

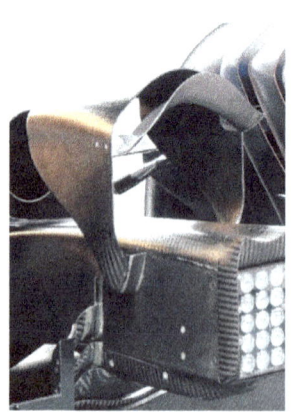

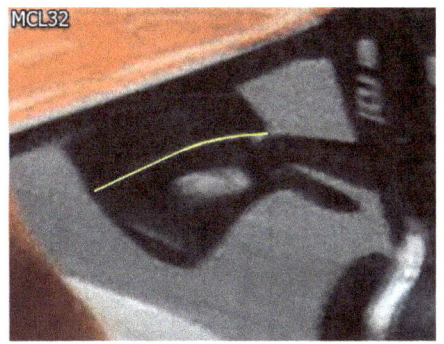

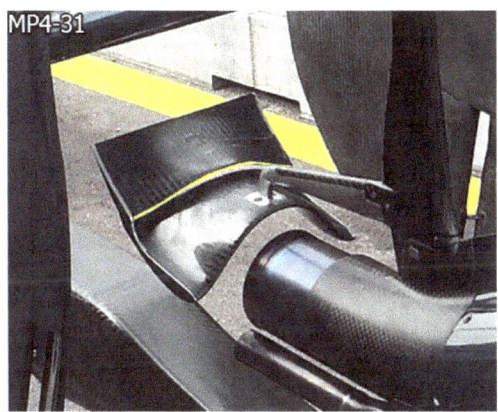

→ Resume:

Angle of Attack	Handling Balance
Increased Front Wing Angle of Attack-	• Decreased understeer. • Increased oversteer.
Decreased Front Wing Angle of Attack-	• Increased understeer. • Decreased oversteer.
Increased Rear Wing Angle of Attack-	• Decreased oversteer. • Increased understeer.
Decreased Rear Wing Angle of Attack-	• Increased oversteer. • Decreased understeer.

www.ingramcontent.com/pod-product-compliance
Lightning Source LLC
Chambersburg PA
CBHW071140220526
45467CB00015B/1606